4-5歲

幼兒全方位
智能開發

英文篇

First Words
英文生字

bread

ball

園丁文化

My Family 1
我的家庭（一）

● Draw a line to match the picture with the correct word.
請用線把圖畫和正確的詞彙連起來。

1.

   •   •   father

2.

   •   •   mother

3.

   •   •   grandfather

4.

   •   •   grandmother

My Family 2
我的家庭（二）

⬤ Fill in the blanks with the missing letters.
請在橫線上填寫正確的英文字母。

1.

s___st___r

2.

___ncle

3.

a___ ___t

4.

br___ther

⬤ Draw a picture of your family.
請把你的家人畫出來。

答案：1. sister 2. uncle 3. aunt 4. brother

3

My Family 3
我的家庭（三）

● Trace the lines to match the words with the correct pictures.
請沿虛線連一連，把詞彙和正確的圖畫配對起來。

daughter

father

son

mother

My Body 1
我的身體（一）

Fill in the boxes with the correct words. 請在空格內填寫正確的詞彙。

shoulder   ear   nose   mouth   eye

1.

2.

3.

4.

5.

答案：1. eye 2. ear 3. mouth 4. nose 5. shoulder

5

My Body 2
我的身體（二）

⬤ Circle the correct words. 請圈出正確的詞彙。

1.
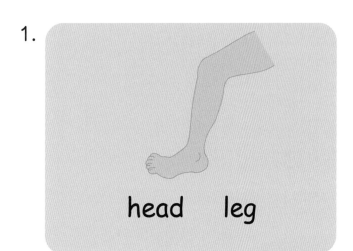

head   leg

2.
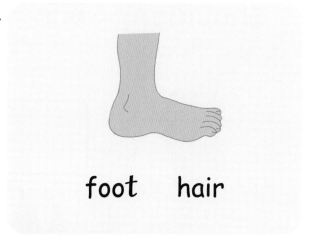

foot   hair

3.

foot   head

4.

leg   hair

Please touch your head and foot.
請碰一碰你的頭和腳。

My Body 3
我的身體（三）

Fill in the blanks with the correct words. 請在橫線上填寫正確的詞彙。

nose　　ears　　hands　　eyes

1.

I can see with my _____ .

2.

I can hear with my _____ .

3.

I can touch with my _____ .

4.

I can smell with my _____ .

Revision 1
複習（一）

● Complete the crossword puzzle. 請完成填字遊戲。

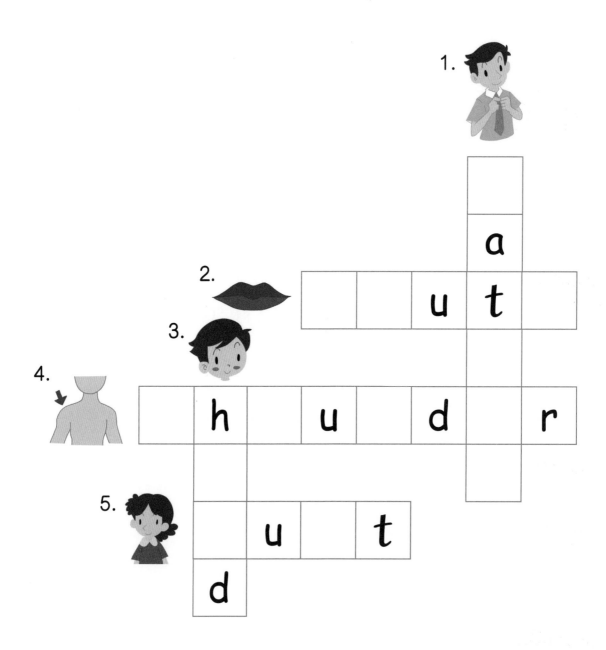

My Classroom 1
我的課室（一）

● Write the words. 齊來寫一寫。

window	window	window

blackboard	blackboard	blackboard

teacher	teacher	teacher

desk	desk	desk

chair	chair	chair

● Draw a line to match the picture with the correct word.
  請用線把圖畫和正確的詞彙連起來。

1.  ● ● glue

2. 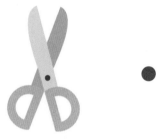 ● ● eraser

3.  ● ● scissors

4.  ● ● crayon

答案：1. crayon 2. scissors 3. glue 4. eraser

My Classroom 3
我的課室（三）

● Fill in the blanks with the missing letters.
請在橫線上填寫正確的英文字母。

1.	There are three __encils.
2.	There is a __uler.
3.	There are two __ooks.
4.	There is a __esk.

My Clothes 1
我的衣物（一）

● Circle the correct words.　請圈出正確的詞彙。

1.	It is a (dress / skirt).
2.	It is a (hat / dress).
3.	It is a (tie / skirt).
4.	It is a (tie / hat).

My Clothes 2
我的衣物（二）

● Put a ✔ in the correct box.　請在正確的 □ 內加 ✔。

1.
　　□ scarf
　　□ pants

2.
　　□ socks
　　□ boots

3.
　　□ pants
　　□ boots

4.
　　□ scarf
　　□ socks

● Draw a picture of your favourite outfit.
　請畫出你喜愛的衣物。

答案：1. scarf　2. boots　3. pants　4. socks

My Clothes 3
我的衣物（三）

● Write the words. 齊來寫一寫。

1.

shirt

2.

vest

3.

shoes

4.

belt

● Circle the correct words.　請圈出正確的詞彙。

1. skirt

shirt

2. boots

books

3. socks

pants

4. hat

scarf

5. ruler

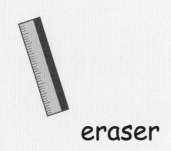

eraser

6. desk

chair

答案：1. skirt 2. books 3. socks 4. hat 5. ruler 6. chair

Food 1
食物（一）

● Draw a line to match the picture with the correct word.
請用線把圖畫和正確的詞彙連起來。

1.

●　　　　●　ice cream

2.

●　　　　●　sausage

3.

●　　　　●　biscuit

4.
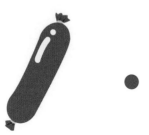
●　　　　●　milk

答案：1. biscuit 2. milk 3. ice cream 4. sausage

16

Food 2
食物（二）

● **Circle the correct words.　請圈出正確的詞彙。**

1.

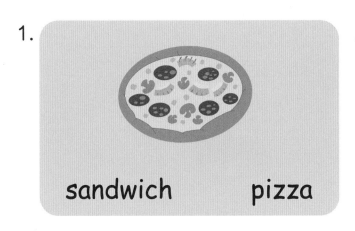

sandwich　　pizza

2.

hamburger　　cake

3.

pizza　　　　cake

4.

hamburger　sandwich

What is your favourite food ?
你喜愛吃什麼食物？

答案：1. pizza 2. hamburger 3. cake 4. sandwich

Food 3
食物（三）

Colour the picture. 請依指示把圖畫填上顏色。

bread

lollipop

egg tart

juice

jam

答案：

18

Food 4
食物（四）

Match the sentences with the correct pictures. Write the correct letter in the ☐.

請把句子和圖畫配對起來，並把代表正確句子的英文字母填在☐內。

A. I like chocolate.   B. I like jellies.

C. I like noodles.   D. I like popcorn.

1.

☐

2.

☐

3.

☐

4.

☐

答案：1.D 2.A 3.B 4.C

Fruits 1
水果（一）

Draw a line to match the picture with the correct word.
請用線把圖畫和正確的詞彙連起來。

1.

　●　　　　　●　**watermelon**

2.

　●　　　　　●　**mango**

3.

　●　　　　　●　**strawberry**

4.

　●　　　　　●　**apple**

答案：1. strawberry 2. apple 3. watermelon 4. mango

Fruits 2
水果（二）

Put a ✔ in the correct box. 請在正確的 ☐ 內加 ✔。

1.
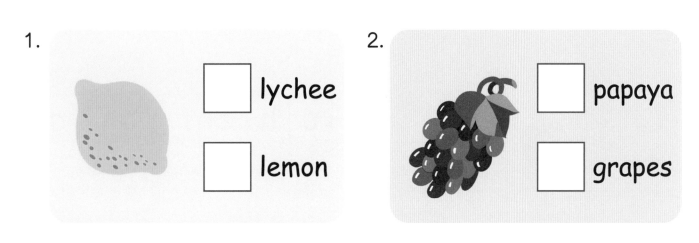

☐ lychee

☐ lemon

2.

☐ papaya

☐ grapes

3.
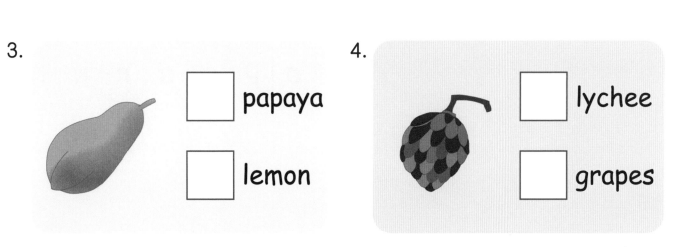

☐ papaya

☐ lemon

4.

☐ lychee

☐ grapes

Please draw a picture of your favourite fruit.
請畫出你喜愛的水果。

答案：1. lemon 2. grapes 3. papaya 4. lychee

Fruits 3
水果（三）

Colour the correct letters of the fruit's name.
請把水果名稱的正確串法填上顏色。

1.

o	c	h	e	r	r	y	a

2.

a	o	p	e	a	r	u	n

3.

b	a	n	a	n	a	q	b

4.

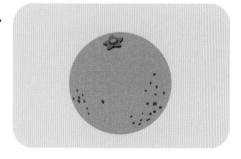

r	e	o	r	a	n	g	e

Fruits 4
水果（四）

Circle the correct words.　請圈出正確的詞彙。

1.
pineapple

kiwifruit

2.
dragon fruit

starfruit

3.
starfruit

kiwifruit

4.
dragon fruit

pineapple

Do you like eating fruits?
你喜歡吃水果嗎？

答案：1. pineapple　2. starfruit　3. kiwifruit　4. dragon fruit

23

做得好！　不錯啊！　仍需加油！

● Complete the crossword puzzle. 請完成填字遊戲。

1.
2.
3.
4.
5.
6.
7.

	t			w	b			y

| u | | e |

| e |

| m | | g | | d |

| g | | h |

| e | a | | e |

Animals 1
動物（一）

● Colour the correct letters of the animal's name.
請把動物名稱的正確串法填上顏色。

1.

f	i	c	r	a	b	d	a

2.

g	r	a	f	i	s	h	e

3.

f	d	o	l	p	h	i	n

4.

o	c	t	o	p	u	s	s

答案：1. crab 2. fish 3. dolphin 4. octopus

Animals 2
動物（二）

● Draw a line to match the picture with the correct word.
請用線把圖畫和正確的詞彙連起來。

1.  ●　　　● **tiger**

2.  ●　　　● **elephant**

3.  ●　　　● **bird**

4.  ●　　　● **giraffe**

Animals 3
動物（三）

● Fill in the blanks with the missing letters.
請在橫線上填寫正確的英文字母。

1.
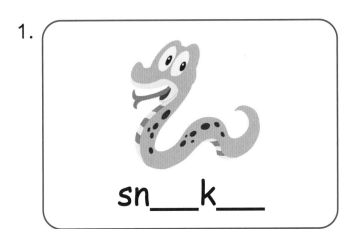
sn___k___

2.

b___tt___rfly

3.
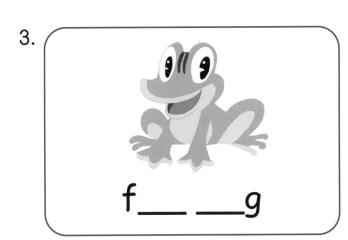
f___ ___g

4.

t___rto___se

● Please draw a picture of your favourite animal.
請畫出你喜愛的動物。

答案：1. snake 2. butterfly 3. frog 4. tortoise

27

Animals 4
動物（四）

● Circle the farmer's animals.　請圈出農夫的動物。

答案：A, E, F, G

Toys 1
玩具（一）

⬤ Circle the correct words.　請圈出正確的詞彙。

1.

toy gun

balloon

2.

drum

robot

3.
robot

balloon

4.
drum

toy gun

⬤ Please design your dream toy and draw a picture of it.
請設計你的夢想玩具，並把它畫出來。

● **Complete the crossword puzzle.** 請完成填字遊戲。

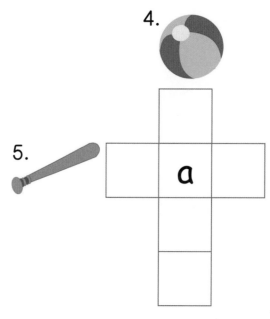

Toys 3
玩具（三）

● Write the words.　齊來寫一寫。

1. 　**car**

2. 　**doll**

3. 　**bear**

4. 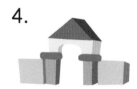　**blocks**

What is your favourite toy ?
你喜歡什麼玩具？

● Draw a line to match the picture with the correct word.
請用線把圖畫和正確的詞彙連起來。

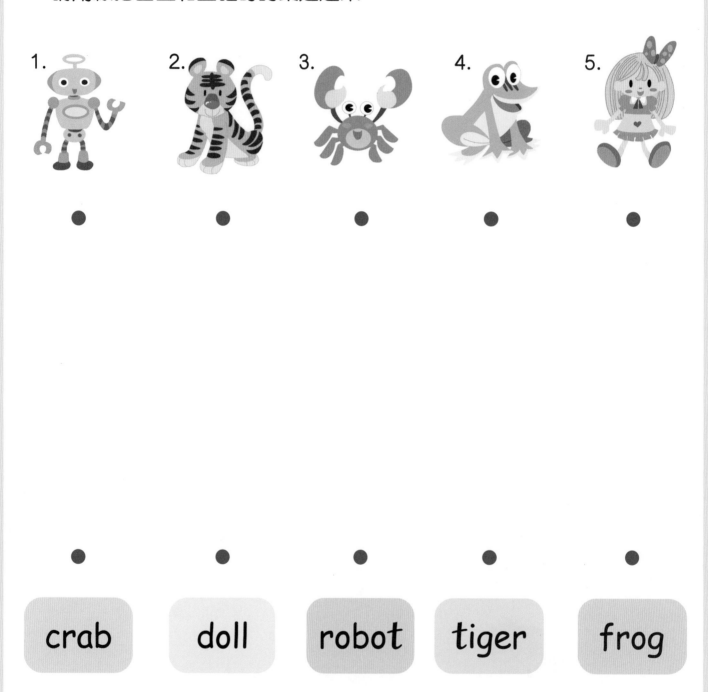

1. 2. 3. 4. 5.

| crab | doll | robot | tiger | frog |